Let's Count Halloween!

A Fun Kids' Counting Book For Children Age 2 to 5

Alina Niemi

All rights reserved. Copyright © 2014 by Alina Niemi.
No part of this book may be copied, retransmitted, reposted, duplicated, or otherwise, by any means, electronic or mechanical, including photocopying, recording, or by any information storage and retrieval system, without permission in writing from the author, except by reviewers who may quote brief excerpts in connection with a review. Any unauthorized copying, reproduction, translation, or distribution of any part of this material without permission by the author is prohibited and against the law.

What do you think of this book?
Please consider leaving a review. It helps me get better and helps others decide if this book will help them. Thank you!

Find more fun books at alinaspencil.com
Plus links to stores with customizable shirts, stickers, mugs, bumper stickers, personalized pet bowls, phone and laptop cases, custom photo wall clocks, magnets, greeting cards, and much more!

ISBN: 978-1-937371-05-0

Alina's Pencil Publishing

2 two shadows

3 three trick-or-treat bags

5 five cauldrons

6 six skeletons

7 seven bats

8 eight tombstones

9 nine toads

11 eleven
jack-o-lanterns

13 thirteen bones

14 fourteen eyes

15 fifteen coffins

15

16 sixteen ghosts

17 seventeen brooms

18 eighteen black cats

19 nineteen witch hats

20 twenty candy corns

Check out the other books in the series!

Let's Count Trucks

Your truck-loving toddler boy or preschooler will have fun and learn to count with this educational book.

Find colorful pictures of pickup trucks, firetrucks, tractors, and more. Free coloring and doodling pages are also included in the print book.

Your child or grandchild can point to each object and count out loud. Many toddlers know their numbers but get confused when counting objects. This is a fun way to practice numbers 1 to 10.

Recommended for children age 2 to 5

Let's Count Summer

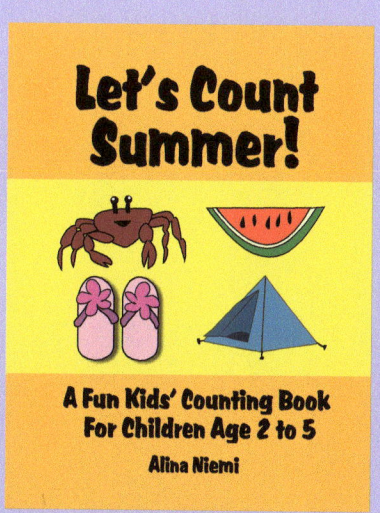

Find favorite items you'll see during the summer months or at the pool or beach, like swimsuits, lemonade, and seashells. Kids will love the cute pictures and can practice counting numbers 1 to 20.

Recommended for children age 2 to 5

Also available: Strange Ice Cream

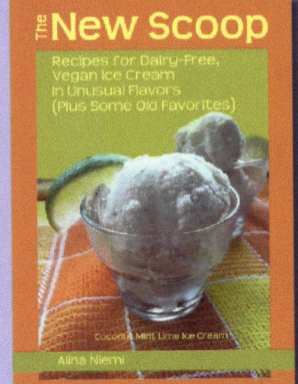

Like ice cream but can't have dairy products? **The New Scoop: Recipes for Dairy-Free, Vegan Ice Cream in Unusual Flavors (Plus Some Old Favorites)** contains recipes for ice cream, sherbet, sorbet, and frozen yogurt, all without milk or eggs.

Try Peanut Butter and Jelly Ice Cream, Cucumber Mint Frozen Yogurt, or Pineapple Sherbet. Tropical flavors include guava, lilikoi (passionfruit), lychee, and mango. Learn to make mochi ice cream and yogurt at home!

Like to draw?

Learn about Hawaii's culture, food, and animals while you doodle. There are over 100 pages for you to color, draw, or decorate.

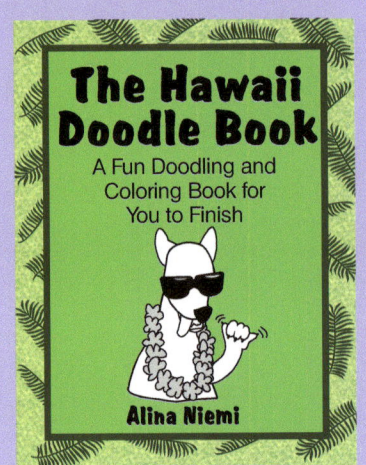

Contains how-to-draw tips, and a glossary, plus a guide to Hawaiian language and pronunciation.

Available 2014:

When Masako's musubi in her lunch turns into a sumo wrestler, she tries to show her friend. But there is only a rice ball. Did she imagine it? Find out in **Masako's Bento Lunchbox Surprise**.

www.alinaspencil.com

www.ingramcontent.com/pod-product-compliance
Lightning Source LLC
Chambersburg PA
CBHW040032050426

42453CB00002B/92